Border Breeds of Sheep

by John Usher

with an introduction by Jackson Chambers

Self Reliance Books

Get more historic titles on animal and stock breeding, gardening and old fashioned skills by visiting us at:

http://selfreliancebooks.blogspot.com/

Introduction

I am pleased to present yet another practical title on breeding and raising livestock.

The work is in the Public Domain and is re-printed here in accordance with Federal Laws.

As with all reprinted books of this age that are intended to perfectly reproduce the original edition, considerable pains and effort had to be undertaken to correct fading and sometimes outright damage to existing proofs of this title. At times, this task is quite monumental, requiring an almost total "rebuilding" of some pages from digital proofs of multiple copies. Despite this, imperfections still sometimes exist in the final proof and may detract from the visual appearance of the text.

I hope you enjoy reading this book as much as I enjoyed making it available to readers again.

Jackson Chambers

Prefatory Note.

Two of the following papers on the "Border Breeds of Sheep" were written about two years ago for *The Field*, and formed part of a series of articles from various pens, published from time to time in the columns of that journal, on the distinct breeds of cattle and sheep in Great Britain. The author wrote them in a fluent —some people may think rather a flippant —style, purposely avoiding dry statistics, with the view to suit the taste of general newspaper readers. Had he composed them with an eye to publication in their present form, he would have been a little more practical. At the request of the publisher—who considers them worthy of reproduction—he has revised them, and added an article on Blackfaced Sheep, thus, as it were, completing the pure breeds of sheep peculiar to Scotland and the Border counties.

The publisher has spared neither trouble nor expense in getting up the small volume in an attractive form ; and should it attain any measure of success, the author considers it will be more owing to the excellence of its illustrations than any intrinsic merit possessed in the articles themselves.

STODRIG, KELSO,
January, 1875.

Contents.

———◆———

Blackfaced Sheep.

Blackfaced Sheep.

————•———

THE early history of the breed, commonly called Highland or Blackfaced, is involved in obscurity. Little doubt can exist that they are a native breed, and we may just say of them, as " Topsy," in Uncle Tom's Cabin, " 'Specks they growed." The probability is, that they were allowed for ages to pick up a precarious subsistence as best they could, and multiply and replenish the earth according to their own natural instincts. Although there is no record of any great improvers, as Bakewell in Leicesters, or Robson in Cheviots, suddenly bringing about a revolution in their history, a great improvement has been effected, more especially if we are to believe that the ancient "Dunfaces" were their original progenitors. No doubt, it was brought about gradually by men of intelligence and judgment, in careful selection, putting together animals of the best type, and breeding from their produce; but as Blackfaced flocks occupied wide tracts of country where fencing was unknown, the benefits of such selection were often in a measure lost by the tups being put to them indiscriminately. This naturally accounts for their improvement being slower than in breeds placed in more favourable circumstances.

In our treatise on the Cheviots, we mention that after the great improvement on the breed effected by Robson of Belford, they in a great measure supplanted the Blackfaces wherever soil and climate were at all suited to their production. Hence, it may be considered somewhat inconsistent when we say that of late years there has been a decided reaction in favour of the Blackfaced Sheep. This is attributable to the disastrous seasons of 1859-60 and 1860-61, which will long live in the memory of hill farmers. A succession of cold, stormy weather entirely killed vegetation, and consequently Cheviot flocks got so reduced in condition that they died in great numbers; while it was observed that the loss in Blackfaced was infinitesimal in proportion. The belief gained ground that the keeping of Cheviots had been carried beyond its natural limits, and the result was that many sheep farmers took to keeping a proportion of Blackfaced on their higher grounds. As Cheviots are usually kept on what is called " green land," whereas Blackfaced have generally a mixture of moss and heather, a controversy arose, and in fact still exists, on the point,

—whether, if the Blackfaced had been confined to the " green land," with all the natural juices dried out of it, they would not have succumbed as readily as the Cheviots ?   Blackfaced breeders generally repudiate this opinion, and we feel disposed to agree with them, being fully persuaded that under any circumstances their favourites are the hardier, less liable to rot, and altogether so tenacious of life that they will live on, though reduced to the smallest pittance and the most abject poverty.

Our own recollections of Blackfaced sheep carry us back nearly half a century, when the father of the writer took a large stock farm, about the highest in Lammermoor, which he stocked entirely with them, and we became associated with him in their management.   We remember well that then Gillespie of Douglas Mill, in Lanarkshire, was the man principally connected with the improvement of the breed, and, in fact, his flock considered at the top of the tree. In drafts from it ours was principally, indeed almost exclusively, selected, and the consequence was that we soon attained a considerable local notoriety, carrying nearly all the prizes for tups and gimmers at Gifford, Lauder, and other local shows for a good many years. We then gathered any little practical experience we ever had in the management of Blackfaced sheep, and from personal observation can still recall many of their peculiarities—their strong affection for some favourite haunt often prompting the ewes to travel long distances to it on the eve of lambing—their wonderful maternal instinct in sticking to their lambs in the most suffocating snow-drifts—their indomitable industry and perseverance in working among the deep snow with their feet, and earning a scanty subsistence in the most severe weather—and their

wonderful powers of endurance under the most
trying circumstances. Among many others of a
similar nature, we remember an instance of a wedder coming out alive, after lying under the snow
twenty-eight days, dreadfully emaciated, but with
constitution sound as ever.   From the singularity of
his history, he was kept on for several years, and
held the honourable position of " Snawbreaker " to
the flock, in leading them to the hill when deeply
covered with snow.   We have also lively reminis-
cences of the chivalrous spirit of the tups at
certain seasons, walking back from each other some
five or six yards a piece, as if by mutual consent, and
then meeting in the centre in full career, like knight
errants in the tourney, their horned foreheads clos-
ing like the clang of armour, and occasionally dis-
locating each other's necks.   To show the wonderful individuality the Blackfaces possess, like " the

human face divine," no two being exactly alike, we may mention that in our highest-lying hirsel of ewes, which was herded for many years by a man of great simplicity of character, who used to be called "a sheep among men, and a man among sheep," he made no scruple in saying that he knew every individual sheep of his flock, consisting of about 700, and thought there was nothing at all wonderful in it. We were among the first to cross some of our ewes with Leicester Tups in so high a a locality, a system which has since become so popular. It had, however, its disadvantages : they came much leaner out of their wool, and, besides, it curtailed our choice of ewe lambs for stock. The ewes are wonderful milkers, as much superior to the Cheviots in this respect as Cheviots to the Leicesters, and their lambs less liable to scour. This was evinced by the fact, that in peculiarly fine warm summers all the lambs not required for stock were sold to the butcher. Our wedders being sold to regular customers in lots of one or two scores, as pot flocks, recalls the following circumstance. In a season of limited supply one of them had accidentally been overlooked, and, instead of wedders, was consequently supplied with a score of young eild ewes or gimmers. He thus inadvertently made the discovery, which had been patent to us long before, that a Blackfaced sheep fattened on its native pasture is the finest mutton in the world, and the competition for them was great in following years.

Within the last twenty years or so, the chief pioneers in the improvement of the Blackfaced breed were John Watson, Nisbet; James Craig of Craigdarrock, Biggar; David Foyer, Knowhead, Campsie; James Greenshiels, Westown, Douglas; Thos.

Aitken, Listonshiels; and John Archibald, Over-shiels, Stow. We might add many more of nearly equal celebrity, the difficulty being to know where to stop; and we feel almost warranted in saying that the latter gentleman has fairly established his pre-eminence, by carrying the highest honours, for several years, at the Highland Society's and other important shows. Mr. Archibald also rents the hill-farm of Midcrosswood, in the Pentland Hills. His son, Mr. James Archibald, has been associated with him in the management of Blackfaced stock since his boyhood, and no man can be more thoroughly enthusiastic and practical than he. At his arable farm of Duddingstone, near South Queensferry, the very finest specimens of the breed may any day be seen, enjoying a little generous treatment preparatory to making their appearance in auction and prize rings.

It is a curious fact that breeders who have attained eminence in the management of any particular class of animals have generally an eye to the lines of beauty and symmetry in others; their success in the one in which they specially excel proceeding often from fortuitous circumstances. We have an instance of this in Mr Archibald and his sons, who, on their farm of Glengelt, on the confines of Lammermoor, have a flock of Cheviots of rare quality, and bid fair to dispute supremacy with Cheviot breeders of older standing. In like manner we find Lord Polwarth, so justly celebrated for his famous flock of Leicesters, showing a bold front in the breeding of Shropshire Downs, and recently selecting a breed of shorthorned cattle, which have already given augury of future fame. We have a more striking example still in Mr Elliot of Hindhope—the acknowledged representative man in Cheviots—whose name we see frequently in prize lists all over the kingdom, in Leicesters and Blackfaces in sheep; shorthorns, kyloes, and crosses in cattle; Clydesdales, hunters, and hackneys in horses; while in the canine species he is equally at home in foxhounds, greyhounds, and collies. It is well known that in the latter his taste and judgment have been in requisition by the highest lady of the land.

Breeders have, of course, their own peculiar tastes; but it is generally allowed that the frame of the Blackfaced sheep should approximate very closely to that of the Cheviot. During the writer's experience, the fashionable taste in horns was somewhat different from the present; they could not be too close to the head, if they only cleared it, and in trying to reach the standard of excellence, there was a constant necessity for the use of the saw. This was an error; the present taste, of the horns being wider set, and rising in a fine circular arch, quite clear of the head, is not only handsomer, but betokens, we think, a more fully developed animal. They should never rise high on the *cantle*, which is not only ungraceful, but has always a most pernicious effect on ewes in lambing. The horns should be hard and flinty, and on no account blood-red, which, besides being unsightly, indicates a softness of constitution. We prefer the colour of the face to be either entirely black, or *brocked*—viz., black and white, clearly defined, without running into grey—and both face and legs ought to be clean, and free from all dunness or tuftiness. The flow of the wool should reach within a few inches of the ground, and be free from dead hairs. Animals of this type possess a great deal of style and quality, and are perhaps more agile than any other of the sheep species.

The localities where Blackfaced sheep may be

said to have attained the greatest perfection in the South of Scotland are Lanarkshire, Ayrshire, Mid-Lothian, Roxburghshire, &c. There they are either mixed ewe and wedder, or entirely ewe or breeding flocks. They are allowed to go at large over their walk, and are gathered or disturbed in any way as seldom as possible. Occasions of this sort are now less frequent than formerly, as weaning is hardly ever resorted to in lambs intended for stock, unless in the case of the draft ewes, for ten days or so, to allow them to dry; washing is given up, and voted an institution which, in their case, does not pay; and smearing also is abandoned for the easier practice of dipping, except in very exposed situations. Like Cheviots, the ewes have their first lambs at two years old, the lambing season commencing about the middle of April, and are sold at five or six, generally to produce a single

crop of half-bred lambs, and then be fattened
for the butcher.   In ordinary years they can tide
over the winter without any auxiliary food at all ;
but in a deep storm, a small quantity of natural hay
is enough to supply their wants.   In the Midland
Counties mixed ewe and wedder flocks may be said
to be the rule, while in the Northern Counties they
are usually wedders entire.   These are generally
bought as lambs, which are sent for wintering into
the lower grounds in the neighbourhood of the sea-
shore, the two and three-year-old wedders being
afterwards wintered on the lower grounds of the
summer grazings, and then sent to the Southern
markets to be sold for feeding off.   In favourable
seasons a good many of them are fair fat from the
hills.

The principal market for ewe hoggs used to be
held at West Linton in May, commencing by grey

day light, and, when much in demand, sales of high-class animals being often effected the night before. It is now removed to Lanark, and there, we are told, sales often commence a day or two before the legitimate market. The market for ewe and wedder lambs is also held at Lanark in August. Ewe hoggs have sometimes reached as high as £46; ewe lambs, £29; and wedder lambs, £21, 10s per clad score. The principal market for north country ewes and wedders is Falkirk. The tups are generally sold by auction, the greatest market for them being Edinburgh, and the prices they bring clearly show that the spirit of improvement is as great in Blackfaces as any other breed. In 1873, Mr. Greenshiels reached the high average of from £13 to £14 for his lot, and Mr. Archibald about £15 on a lot of sixty. The improvement in the quality of wool has also been considerable; but as it is a secondary matter in the culture of the breed—a good covering from the cold being the main point—it has perhaps scarcely been commensurate with that in other respects. The average weight of fleece over a flock may run from 3lbs. to 5lbs., the latter being an extreme weight. The intrinsic value of the Blackfaced breed lies in the superior quality of mutton, and the excellent cross they make with the Border Leicester tup. This cross is not much slower in coming to maturity than the Leicester Cheviots, more especially—judging from our own experience—if they are pushed on with extra feeding on grass in summer, and their mutton is not very much inferior to Blackfaced entire.

In reference to the circumstance previously mentioned, of hill farmers having lately increased the breeding of Blackfaced sheep on their high-lying lands, it is a curious fact that this can only be done with advantage by fencing them off from the Cheviots. If allowed to graze indiscriminately, they are far too active and industrious for their more luxurious neighbours. Starting from their heights with the dawn, they sweep down upon the lower grounds, nipping the tender herbage with the dew on it; then make the circuit of the middle walk, and are first up again in the evening to their favourite haunts. Thus the Cheviots may be said to be much in the same plight as anglers following in the wake of more skilful lovers of the gentle craft down the tempting streams and whirling eddies of some well-frequented river.

On the whole, we may safely conclude that the Blackfaced sheep will maintain their prestige as a most interesting and useful auxiliary in hill farming; and, where they are confined to their legitimate limits, pay for judicious breeding and careful management, as well as any other. Long may their lithe and agile figures, which artists are so fond of depicting on their canvas, dot old Scotland's rugged mountain sides, adding to their picturesque beauty; and long may their more fully-developed forms grace the noble parks of our aristocracy, even in the heart of merry England !

Cheviot Sheep.

Cheviot Sheep.

———•———

THE Cheviots—a range of hills in the Border counties of England and Scotland—were the early home of the Cheviot sheep, whence they derive their name, and to which they were exclusively confined for many generations. They seem to have been a native breed, although a legend still gains credence, especially among shepherds, that the first of them were imported into the country by the Spanish Armada, having swam to land from some of the shipwrecked vessels of that ill-fated expedition that were drifted on the Western Isles. They are generally described as small sheep, very light in bone and wool, with brownish heads and legs, and hardy constitution; their scraggy frames bearing very little resemblance to the well-proportioned Cheviots of the present day. Nevertheless, from their adaptation to the soil and climate, they appear to have spread over a great part of the elevated lands in the south of Scotland long before an attempt was made to improve them. The earliest recorded attempt was about a hundred years ago, and was eminently successful. The merit of this is universally accorded to Mr. Robson of Belford, although Cheviot breeders of the present day differ materially regarding the cross he made use of. We have it from Mr. Robson Scott of Newton—a grandson of Mr. Robson—that he travelled over the greater part of England for the purpose of seeing various breeds of sheep in different districts, with the view of selecting rams to cross his flock of Cheviots. The sheep he considered most suitable were of a breed then existing in Lincolnshire, of which he purchased several rams to put to selected ewes. The cross answered admirably, greatly improving the flock in every respect, without materially lessening its hardy character. Mr. Robson then occupied several high and stormy farms on the Border, and the cross breed throve well upon them. Twenty years afterwards he made a second visit to Lincolnshire to obtain another infusion of the same blood, but found the breed had become so much larger and less hardy that he declined to venture on them. The theory of Mr. Aitchison of Linhope, a high authority in Cheviots, as well as other eminent breeders, is that the breed Mr. Robson imported were Bakewell's Leicesters, with which he crossed a few select Cheviot ewes, and that the offspring of this cross were sent to the hills to cover his extensive flocks. The great resemblance between the two breeds raises a strong presumption in favour of this hypothesis ; but, on the other hand, the tenderness of the Leicesters makes it very improbable that such a cross could stand the winters of so stormy a climate. We have besides, in later times, been cognisant of instances where a slight dash of the Leicester blood was introduced, and proved detrimental to the hardiness of the breed, and experimenters were generally fain to retrace their steps. Of the two assertions, therefore, we incline to that of Mr. Robson Scott, more especially as it is not merely derived from tradition, but, as he solemnly affirms, from an oral statement he had from his grandfather. Under any circumstances, Mr. Robson stands confessed the great improver of the breed, although, like Bakewell in Leicesters, the means he used are involved in some obscurity. This early cross gave a correctness of form and symmetry that has never yet been surpassed : greater bone has no doubt been introduced in the present day ; but, in the opinion of many Cheviot breeders, to an unprofitable extent, as greater bone often implies reduced numbers.

Mr. Robson's flock thus proved the nucleus from which Cheviot breeders drew their supply of rams for many years. His mode of selling is said to have been somewhat unique. A ticket was attached to each sheep with the price put on him, so that customers could choose according to their tastes and means. The impetus given to the breeding of Cheviots was immense : they rapidly found their way into other districts of Scotland and the north of England, supplanting the blackfaced breed, which, like the aborigines in India and America, may be said to retire before the advancing wave of civilization.   Let it not be supposed, however, that we disparage the blackfaced breed of sheep.   For hardiness and beauty they are unsurpassed, and still yield a profitable return in regions where Cheviots could not live.   Our earliest associations in sheep farming are connected with them; and we well remember a severe snow-storm in Lammermoor, late in April of 1827, when the Cheviot ewes, losing the instinct of maternal affection, left their newly-dropped lambs to perish in scores, while the blackfaced stuck closely to theirs, and the loss in them was a mere trifle.   Our memory still clings to their black and mottled faces, bright eyes, and beautifully arched horns, with all the freshness of a first love.

Early in the present century the Cheviot sheep were largely introduced into the northern counties of Scotland, chiefly by farmers of large capital on the Borders. Numbers of small crofters were turned out of their holdings, which were changed into extensive sheep walks. There can be no doubt that the movement, although unpopular at the time, was the means of increasing production, and proved in every case of judicious management a most profitable investment.

In later times the condition of Cheviot flocks has been greatly ameliorated by draining, shelter, providing a plentiful supply of food for use in stormy weather, and other modern improvements. Mr. Aitchison of Linhope may be said to have been the pioneer both in the advocacy and practice of the system of cutting a considerable quantity of hay, not only on the open grounds, wherever the deepness of the soil afforded an extra covering, but by having several enclosures on each farm where hay could be produced sufficient for its requirements, thus making them self-sustaining. These inclosures are also useful as a run for the weaker ewes and lambs, and afford an early bite, so essential to ewes in the lambing season. To use Mr. Aitchison's own forcible language: "Hay is the sheet-anchor of the stock farmer." We doubt not some of our readers will recognise in Mr. Aitchison a man not only intimately associated with the improvement of Cheviot stock, but of agriculture in general, and recall with a thrill of pleasure his deep-toned voice, clear enunciation, and fervid eloquence in returning thanks at the banquets of the Highland Society of Scotland for the toast of "The Tenantry," or the halo of romance he threw over his subject when, in language rivalling in sublimity the poetry of "Ossian," he proposed "The Peasantry of Scotland."

The practical management of a Cheviot flock is, on the whole, exceedingly simple. Generally speaking, they go at large over the farm during the whole season, individual sheep never taking a very wide range. The area required for each varies from about two to four acres, according to quality. In some cases the hoggs are kept separate from the ewes, which gives an opportunity of supplying them with more generous treatment in stormy weather; but frequently they are allowed shortly after weaning to graze together. This gives them the advantage of a mother's care, for they generally recognise each other. In some cases they are allowed to go on without being weaned at all; but we think such a system must be injurious to the future progeny. Ewes have their first lambs in April at two years old, and are sold as draughts at five or six, being replaced by the best of the ewe lambs. They are invariably sold for producing a crop of lambs by Leicester tups. These, with the wedder lambs, the small ewe lambs, and wool, usually form the whole produce of the farm. This applies to Cheviots in the southern counties of Scotland; in the north the practice differs considerably. There the wedder lambs are not sold, but kept on till sold as wedders at three years old. The wedder hoggs are never wintered at home, but sent into winter quarters in Ross-shire and neighbouring counties—some as far as Aberdeenshire—where they have the *outrun*, as it is called, on arable farms—viz., nearly the whole grass—on which they are kept till the weather becomes stormy, when they are folded on turnips. They are sent about October 10, and remain till the beginning of April. The cost of wintering, including smearing (which operation takes place shortly after reaching their winter quarters), varies from about 7s to 9s each.

With the exception of the great Inverness market

in July, where large sales are made by *character*, for delivery later in the season, markets for Cheviots are held in autumn—the most important being Melrose and Lockerbie for lambs, and Falkirk September and October trysts for ewes and wedders. Besides these, auction marts have sprung up in various quarters, where large quantities are disposed of; and although we question the policy of selling stock in bulk in this way, thus superseding old established markets, and paying 3d and 4d in the pound for doing what farmers ought to be able to do for themselves, there can be no doubt that for the sale of single sheep it is admirably fitted. Mr. Aitchison was the first to introduce the system of selling Cheviot tups by auction at his farm of Menzion, in Peeblesshire, more than forty years ago; the practice is now universal. One of the most attractive sales of the season is held at Beattock.

Mr. Bryden of Kennelhead (late of Moodlaw), long known as a most successful breeder ; the Messrs Carruthers of Kirkhill ; and Mr. Johnstone of Capplegill—almost equally celebrated—have an annual sale there, and draw purchasers from all parts of Scotland. Last year about 120 tups averaged 10 guineas each. Similar sales are held in various localities, one of the most important being at Hawick in September, where, among others, the lots of Mr. Aitchison, and Mr Elliot of Hindhope, always command a large attendance and a deal of spirited bidding. The latter gentleman has for many years been a most successful exhibitor of Cheviots at Highland Society's and other shows, carrying everything before him. We cannot resist giving an anecdote, which shows that his fame as a breeder must even have reached the ear of royalty. Happening to be an exhibitor at the Smithfield Show,

Mr. Elliot took the opportunity of visiting the Home Farm at Windsor, when he had the honour of being commanded to wait upon the Queen. Her Majesty, with that graceful condescension for which she is remarkable, received him with a cordial shake of the hand, desired him to be seated, and entered freely into conversation with him. While Mr. Elliot may well be proud of such an honour, we doubt not that Her Majesty also was gratified by the interview, and thought him both in appearance and intelligence an admirable type of the Scottish Borderer.

Harking back from this digression to our subject, there is perhaps no finer animal of the sheep species than the Cheviot tup. Possessing the general conformation of the Border Leicester, he is altogether a more stylish sheep, carrying his head higher, with greater fire in his eye and grace in his movement. Compared with the Leicester, he is as a cavalier to an alderman.

Besides reproducing their own kind, the Cheviots are valuable for crossing with the Border Leicesters; the former giving hardiness, the latter greater tendency to fatten. By infusing the two breeds in different proportions, other breeding stocks are raised, suited to medium soils and temperatures. Thus, taking the Leicesters as the centre of agricultural improvement, the other may be said to radiate. First, we find three-parts-bred in the intermediate; next, half-bred in the higher altitudes; then we come to Cheviot entire on their native mountains; and above and beyond them, our old favourites the blackfaced, among their fastnesses of rock and purple heather.

Cheviot sheep are seldom shorn before July, the weight and fineness of the fleece depending on the nature of the pasturage; the texture being finer on dry, sweet herbage than on coarse grass, and bringing a higher price. It has a steadier demand than almost any other, being extensively employed in the manufacture of tweeds, now so commonly used in clothing, from the prince to the peasant. Coming down from the poetry, so associated with the Cheviots in the lights and shadows of pastoral life, to the inevitable prose (for to mutton they must all come in the end), that of the Cheviot sheep may fairly be put down as one of the luxuries of life. It has always been a nice point whether this or the blackfaced is the finer, and we recall an incident which occurred many years ago, in which the father of the present writer bore a part. He was a great enthusiast in blackfaced sheep, and having the honour to be a special favourite with Sir Walter Scott, and an occasional guest at his table, begged his acceptance of a few wedders to convince him of the superiority of the blackfaced mutton to the

Cheviot, of which Sir Walter was in the habit of keeping what is called in Scotland a pot-flock. Sir Walter accepted them on condition that he would dine with him, along with a few friends, to test their respective merits, when a saddle of each should be presented, having received the same advantage of the culinary art. The verdict was in favour of the Cheviot, to the infinite delight of the great poet and novelist. Dissenting, however, from this judgment, we venture to remark that the quality of both depends very much on the feeding. For delicacy of flavour, we never tasted any mutton equal to that of a *yield* young ewe or gimmer of either breed that happened to get fat on its native pasture.

In taking leave of the subject, it may be stated without fear of contradiction that no animal has conduced so much to the prosperity of the Scottish farmer as the Cheviot sheep, and more especially to those who have engaged exclusively in hill farming. This may be partly attributable to the fact that stock farming is generally embarked in by men of capital, as it involves a considerable immediate outlay, and, the farms being usually large, competition for them is necessarily limited; whereas, arable farms are competed for by men who have made money in other walks of life, and, the demand being greater than the supply, rents have in many cases become exorbitant. Stock farmers are, besides, not nearly so much influenced by the weather, and their expenses are nothing in comparison. The practised working of the stock farm is managed by a few shepherds—a class of men in the rural districts of Scotland distinguished for great moral worth and simplicity of character. They receive their wages in the grazing of one or more cows and a certain number of sheep. They are thus small capitalists, and their interests are identical with their masters'. In arable farming, a very serious increase has arisen in the expenses of cultivation, not only by the rise in wages of agricultural labourers, but in implements, machinery, and, in fact, every department of skilled labour connected with the farm. Thus in 1872, which the elements and other circumstances combined to make to the arable farmer one of the most disastrous seasons on record, to the stock farmer it proved one of almost unexampled prosperity; and, notwithstanding, that 1873 and 1874 have brought on improvements in the price of cereals, and a depression in that of stock, the prospects of the latter will still bear a favourable comparison.

Border Leicester Sheep.

Border Leicester Sheep.

———————

IN tracing the origin of the breed of sheep now commonly called Border Leicesters, it seems almost a work of supererogation to prove that they are descended from a flock known as the Bakewell or Dishley breed ; and the more directly their lineage can be traced to that flock, and their exemption from the introduction of any other strain proved, the more they are generally allowed to be distinguished by symmetry of frame and purity of blood. The breed owed its existence to Mr Robert Bakewell of Dishley, in Leicestershire. By a course of systematic experiments, commenced about the year 1755, in crossing the old Leicesters—said to have been " large coarse animals, with an abundance of fleece and a fair disposition to fatten"—with other long-woolled breeds, probably possessing smaller frames and more symmetrical proportions, he in the course of years worked them into a new breed. As the breeds he used, and the proportions in which he used them, are conjectural (his system having been carried on with much mystery), it seems vain to attempt to enumerate them. Bakewell must have had a good knowledge of animal physiology, and as his aim appears to have been not so much to produce sheep of large size as of fine frame and great aptitude to fatten, it is probable that he connected together animals of the purest blood, nearly allied to one another, thus producing sires which, in their turn, exerted a preponderating influence on their progeny.

That he ultimately succeeded in establishing a distinct breed—their distinguishing feature being a capability of producing, compared with other breeds, the greatest quantity of fat with the smallest consumption of food in the shortest time—is an acknowledged fact. About the year 1760, Bakewell commenced letting his rams for the season at something like a sovereign each; but so rapidly did their reputation increase that in little more than twenty years they had risen about 100 per cent., and in a few years more the demand or mania for the breed was such that seemingly fabulous prices are said to have been realised—as much as £1000 for the season for a single sheep. They thus gradually found their way into other localities; the first draft of them into the Border counties being introduced by the Messrs Cully, who migrated thither from the county of Durham in 1767. The immediate followers of the Messrs Cully were Messrs Smith, Marldown; Thomson, Chillingham; Jobson, Chillingham Newtown; Robertson of Ladykirk; Smith, Learmonth; Compton, New Learmonth; Smith, Norham; Riddell, Timpendean, &c.

Whether some of the early breeders of Leicesters in the Border counties, in imitation of Bakewell's system, tried still further to improve them by crossing in with the Cheviot, a breed possessing fine style and quality; whether the change in their general appearance is due to selecting animals of the pure breed, high on the leg, with white faces and clean bone; and whether the soil and climate have had their influence, are questions that may never be satisfactorily answered. Certain it is that the distinguishing features of the Yorkshire and Border Leicesters, though sprung from the same source,

have diverged considerably ; the former now showing a blueness in their faces and a tuftiness in their
legs, while the latter are white and clean in both,
and more what are generally called *upstanding* sheep.
As the Bakewell breed in early times are described
as having white faces and legs, we leave readers to
draw their own inference. Our hypothesis, that
the Cheviot may have been used by the early
breeders, is suggested by our having seen, within
these few years, a lot of tups bought as pure Leicesters, which, we happened to know, were only
the third cross from a very fine specimen of the
Cheviot tup. The said sheep showed a style and
conformation rarely equalled, and were particularly
good in their necks and heads. Our opinion, however, is, that the flocks tracing the closest lineal
descent from the Dishley, untainted by any other
strain of blood, selected and crossed with taste and

judgment, tended with care, and "all appliances
and means to boot," are still the best in the Border
district. When so bred, they possess the following
conformation :—The head of fair size, with profile
slightly acquiline tapering to the muzzle, but with
strength of jaw and wide nostril; the eyes full and
bright, showing both docility and courage; the ears
of fair size, and well set; the neck thick at the
base, with good neck vein, and tapering gracefully
to where it joins the head, which should stand well
up; the chest broad, deep, and well forward, de-
scending from the neck in a perpendicular line; the
shoulders broad and open, but showing no coarse
points; from where the neck and shoulders join, to
the rump, should describe a straight line, the latter
being fully developed; in both arms and thighs the
flesh well let down to the knees and hocks; the
ribs well sprung from the back-bone in a fine cir-

cular arch, and more distinguished by width than depth, showing a tendency to carry the mutton high, and with belly straight, significant of small offal; the legs straight, with a fair amount of bone, clean and fine, free from any tuftiness of wool, and of a uniform whiteness with the face and ears. They ought to be well clad all over, the belly not excepted, with wool of a medium texture, with an open *pirl*, as it is called, towards the end. In handling, the bones should be all covered, and particularly along the back and quarters (which should be lengthy) there should be a uniform covering of flesh, not pulpy, but firm and muscular. The wool, especially on the ribs, should fill the hand well. When the above conformation is attained, the animal generally moves with a graceful and elastic step, which, in the Leicester sheep, as well as in the human species, constitutes "the poetry of

D

motion," and without which animals, even of high class in any breed, cannot now attain the chief honours in the showyard.

The above may not suit the taste of all Leicester breeders. There has been a tendency in later times to attempt to improve the breed by crossing with sheep of looser frame, and wool of an opener and stronger staple. Such attempts have generally ended in failure, the strain of blood producing tender heads, weak necks and loins, and lack of constitution, and taking many years of careful and judicious management to eradicate. Our opinion is, that in all such attempts, the coarseness, if any, should be on the dam's side, and that the sire should invariably be of symmetrical form and pure blood ; nay, more, we think that where an apparent increase in the weight of fleece and frame has been attained, it frequently proves fallacious when brought to the test of the scales, the extra open fleece weighing lighter than that of a medium texture, and the larger and looser frame, when stripped of the offal, than the more compact, on the same principle as the bone of the thorough-bred horse exceeds in specific gravity the porous bone of the Clydesdale.

There is nothing in the general feeding and management of the Border Leicesters differing materially from that of other breeds. They require good land and good shelter, and, having these, will live and thrive on a small quantity of food. Having a strong tendency to fatten, they arrive at early maturity, and are capable of producing a greater quantity of wool and mutton in a given time than almost any other breed. Their mutton, however, does not stand high in mercantile value, being coarse in the grain and tallowy in the fat. Time was when it found a ready market among the pit-men in collieries. A story is still extant of one of them, when purchasing a portion of a cast ewe with several inches of fat on the rib, on being asked if it was not too fat for him, exclaimed—"Fat! I carena if it war as fat as atween Newcastle and the Scottish Border." Time has brought its changes to the pitman, as to other members of society; higher wages and more leisure enable him to participate in the growing luxuries of the age, and while on gala day he relishes his leg of Cheviot or Southdown, his old tit-bit is only fit for melting into tallow.

The worth of the Leicester sheep does not, however, depend on its value as mutton. In all well-bred flocks the great bulk of the lambs on the male side are kept for tups, and, in like manner, the tops on the female side for breeding purposes. Thus only a limited portion of each, besides the cast ewes and tups of a certain age, find their way into the butcher market. Their intrinsic value consists in their crossing profitably with the Cheviot, Blackfaced, Southdown, &c. The latter are not cultivated extensively in Scotland or the Border counties, being generally considered too tender for the climate. The cross with the Blackfaced makes fine sheep at two years old, yielding mutton of fine flavour. That with the Cheviot also comes to fair maturity at the same age, getting to great weight with mutton of good quality. This cross also forms the foundation for another by breeding from half-bred ewes with the Leicester tup, and producing what are called the three-parts-bred sheep. For this purpose all the tops of the half-bred ewe lambs are kept, and command a higher price than any other. On most lands of fair average quality, where a portion of turnips can be grown, half-bred ewes are kept.   Their produce being a cross nearer the Leicester, their development is rapid ; they are generally forced forward for the butcher market at one year old, or little over ; and, in fact, form the great bulk of the mutton that now feeds our teeming population.   Early maturity and quick returns are the order of the day; epicures in the middle and upper classes are fain to gratify their dainty appetites with mutton of two and three years old ; while Southdown, Cheviot, and Blackfaced wedders of four and five years, with the beautiful West Highland kyloe of similar age, are rarely found, unless in noblemen's and gentlemen's parks, where they are kept, regardless of profit, to tickle the palates of the aristocracy.

Flocks of pure-bred Leicesters are now not confined to the Border counties, but have found their way, wherever soil and climate suit their profitable cultivation, throughout Scotland, even to the "far north," and auction marts for the sale of tups exist in many localities. In Caithness, Sir George Dunbar has, by dint of high farming and selecting sires with great care and regardless of expense from the crack lots of the Border, raised a flock of rare excellence, and his annual sale of tups has reached a very high average. Edinburgh, for numbers, now treads closely on the heels of Kelso. There, the lots of Messrs Clark (Oldhamstocks), Smith (Whittinghame), Smith (Castlemains), Melville (Bonnington), and Lees (Marvingston) find the largest share of popular favour. To our eye, the latter gentleman's stamp of sheep is about the *beau-ideal* of a Border Leicester, combining fair size with fine style and quality. His flock was sold off last year in consequence of farming arrangements; but we are glad to learn that he is about to form the nucleus of another, as no man is better qualified. For sheep of first-class quality, however, Kelso still bears the palm. There, each September brings together upwards of 2000, and merchants from all parts of the United Kingdom. The position of the lots in four auction rings is arranged by ballot, and four auctioneers simultaneously sell single sheep at the rate of one in the minute for more than seven hours. The highest rate is generally attained by Mr Penny, who sells about seventy in the hour, including stoppages and concise preliminary remarks, and finishes in his strong vernacular with a voice as clear as a bell. Lord Polwarth's (Mertoun), Miss Stark's (Mellendean), Rev. R. W. Bosanquet's (Rock), and other crack lots always hold a levee, and thin other rings during their sale. The bidding seldom flags, there being customers for all sorts—tup breeders taking the choicest specimens—

breeders of half and three-parts-bred stock choosing sheep of large frame and open wool; of the black-faced cross, those with closer skins; fat lamb breeders, sheep of good quality, though lacking wool below; while some, contented with a quadruped, if only cheap—alas for them!—fight it out with the butcher and local dealer. It is interesting to the close observer to note the change in the various lots from year to year—some from a bad cross, or untoward local circumstances, losing caste, while others come to the fore, showing the great difficulty in keeping the character of a flock at a uniform standard. We notice in recent years a marked improvement in the lots of Messrs Forster (Ellingham), Laing (Wark), Cunningham (Grahamslaw), Nisbet (of Lamben), Hardie (Harrietfield), Hogarth (Eccles Tofts), &c.; but, most of all, in that of Mr Bell (Linton), his sheep having greatly increased in substance, while still retaining their characteristic gaiety. Mr Torrance's (of Sisterpath) flock also deserve special notice, having, after a partial eclipse, fully attained their former prestige. At the last Kelso Tup Sale perhaps no lot of the same number ever went through the auction ring more uniform in price and general conformation.

Of all the Border flocks, there is none that has maintained such a uniform character as Lord Polwarth's, which deserves more than a mere passing notice. In 1872, his lot of tups never showed to more advantage, the highest-priced sheep reaching £170, and the average about £37, which had only once been exceeded in their history. The flock was formed about the beginning of the present century, being selected from the most direct followers of Bakewell. Our first recollections date back as far as 1835, when Tom Small, of immortal memory, was the presiding genius in their management, and no lover was ever more jealous of the honour of his mistress than Tom of his pet flock. We happened to know him well, and how he spurned the idea of using any strain of blood not strictly Bakewell, and well he could trace them till we got lost in a maze of g.g.g.g.g.g. grand-sires and dams. When Tom felt the infirmities of age creeping on, he was deeply solicitous as to how the flock was to be maintained in its purity, and, ere he " gave commandment concerning his bones," suggested his successor. His choice fell on Andrew Paterson, who had previously been instrumental in bringing a neighbouring flock into a state of great perfection, and, from personal knowledge, Tom knew that " he had the root of the matter in him." Andrew has amply justified his confidence, the success of the flock having been, in the sixteen years he has had

them in charge, not merely uniform, but progressive.
Much conjecture exists as to how the perfection of
the flock is kept up, and as no one ever hears of
Lord Polwarth giving a long price for a tup, it is
generally surmised that there must be a good deal
of in-and-in, or what is called in Scotland *sib*,
breeding.   We had lately an opportunity of seeing
the ewes and gimmers.   Their beautiful blood-like
heads, deep chests, straight backs and bellies, uniform coating of wool, and family likeness, was a
treat to look at.   We fancied that we got some
slight insight into the system of breeding, although
Andrew, like Bakewell, is somewhat mysterious.
Let it, however, be understood that our views are
theoretic.   From the circumstance of the ewes not
being drafted at four or five years old, like the
majority of flocks, but kept occasionally, if good
breeders, till they enter their teens, it is evident that

an opportunity is afforded, with Andrew's profound knowledge of pedigree (of which he is a walking dictionary), to preserve several distinct strains of blood, crossing them from time to time. A good strain is never lost sight of; if rare, it is cherished as a miser would his gold, and animals of rare excellence are never parted with, without leaving their representatives. The procreative powers of nature are never taxed beyond certain limits, and not an ounce of muscular or physical energy is wasted. They are said to be *sib* bred; be it so. The student of animal physiology knows well it is the way to gain symmetry of form, and so long as they keep up fair size and robust constitution along with it, we hold it to be the grand secret of their excellence, the accumulation of one blood, and that blood the purest, enabling them to make their mark wherever they are used, which is as palpable to the eye of a judge, as the cross with a sheep of a totally different breed.

The above remarks on the Mertoun flock were penned more than a year and a half ago, and have been fully borne out by public opinion at the Kelso tup sale of 1873, their highest-priced sheep bringing £195, and average £44, 15s 2½d, the highest they have ever attained. We have also had the opportunity of attending the Hull show in 1873, where Border Leicesters were acknowledged as a distinct breed for the first time by the Royal Agricultural Society of England, and premiums were awarded to them accordingly. Being the first show under the new *regime*, they were not very strongly represented, if we except Mr Forster of Ellingham's famous tup, which is certainly as fine a specimen of the breed as we have yet seen, or may ever see again. Comparing them with other south country breeds exhibited, we were more than ever confirmed in our opinion that, as a rule, the chief defects in the Border Leicesters are in their necks lacking muscular development, and their heads not standing well enough up. Even in this particular, however, they contrasted favourably with the Yorkshire Leicesters, as they have no necks at all—their heads having the appearance of being stuck on their shoulders.

One word on the custom of over-feeding tups so prevalent in the Border counties. It is a great waste in many ways. Sheep so fed cannot be so active under any circumstances, and when taken, as they generally are, to a poorer soil and less genial climate, with the extra feeding entirely suspended, often succumb altogether under such barbarous treatment; at all events, their vital powers are weakened, and, instead of lasting two or three years, they get worn out in one or two. If ewes require to be in an improving condition during conception, why should the sire be in a declining one? and is he likely in such circumstances to impart a healthy constitution to his progeny? We hope to see the introduction of a more natural and healthy system.

Appendix.

APPENDIX.

---◆---

William Aitchison, Esq. of Brieryhill.

It will not be out of place here to give a notice of William
Aitchison, Esq. of Brieryhill, but better known as William
Aitchison of Linhope. Mr. Aitchison at his death, which
took place at Brieryhill on the 4th of April, 1873, had
attained the age of 76, having been born at Linhope in the
year 1797. His father was noted for his knowledge, admiration, and improvement of the Cheviot breed of sheep, qualities which were conspicuously inherited by his son. During
schoolhood, young William did not give much indication of
the possession of brilliant parts, but he was blessed with a
remarkably retentive and ready memory. He had a severe
illness when a boy, which slightly retarded his studies, but he
did not leave school at an unusually early age, and passed two
sessions at the Edinburgh College. He early imbibed a taste
for pastoral farming, and in 1819 had committed to his charge
the extensive farm of Menzion, in Peeblesshire, of which his
father and himself had obtained a joint-lease. The adverse
experiences which fell to his lot through climatic causes might

have dampened the ardour and daunted the courage of many men; but they seemed just what was needed to nerve and brace the youth to contend with hope and perseverance against difficulties. During his lengthened occupation of this farm his wise and well-directed experiments in the improvement of the Cheviot breed of sheep proved most successful, and his flock became widely known and attracted keen purchasers to the periodical auction sales which he was the first to institute. After a time he removed to Linhope, where he continued his efforts to carry the Cheviot to a greater degree of perfection. His high opinion of the breed, his quick perception and discrimination of their points of excellence, and his faculty of adapting means to ends, enabled him to achieve a success which not only rewarded himself but enriched the nation. A greater weight of mutton and a more copious and better staple of wool, with less consumption of food, resulted from these efforts at improvement, while in other points the Cheviot was made a finer and handsomer sheep than before. These were, to use his own phrase, among the " victories of peace" which he claimed as the distinguishing glories of his profession. The qualities of perception, penetration, prudence, and perseverance possessed by Mr. Aitchison gave him great capacity for work, and he at one time held as tenant, in addition to Brieryhill and Glengerry, which he had acquired by purchase, the farms of Stellshaw (in Cumberland), Penchrise, Linhope, and Menzion. The fame of his flock was noised abroad over all the land, and on many occasions his sheep carried the highest honours at the Highland and Agricultural Society's Shows. Though latterly he did not compete for exhibition honours, there was no declension of merit, nor did the market value of his sheep suffer any depreciation. As a pastoral farmer Mr. Aitchison was eminently successful, not simply in regard to the fruits of his labour which he gathered personally, but viewed in relation to the increased productiveness of food and wool which resulted from improvements which he was greatly instrumental in effecting.

Mr. Aitchison was not a one-talented man: nature had been generous to him in her endowments. He early imbibed literary tastes, and his reading and memory, combined with his naturally powerful intellect, enabled him to overcome in great part what might be considered his deficient school training, and made him a fit and favourite companion for the Ettrick Shepherd, Christopher North, and other men of note, with whom he was on terms of intimate friendship. Though making no mark in literature with his pen beyond writing occasional articles, most of which appeared in the *Dumfries Courier*, he took high rank as a public speaker. He had a fine presence, a sonorous and musical voice, a well-stored mind, and a native eloquence which never failed to arrest and hold the attention, and his speech in reply to the toast of the tenantry of Scotland at the Highland and Agricultural Society's dinner at Dumfries in 1837 is described as having astonished and electrified his audience, including many members of the nobility and landlord class. Again, in an im-

promptu speech at a like meeting at Dumfries in 1845, in
reply to the toast of the tenantry, he increased his reputation
as a speaker, and, notwithstanding that he gave utterance to
some views on the subject of the Corn Laws—of which, though
a Conservative, he did not approve—he was most enthusiasti-
cally received.   He distinguished himself by his public ad-
dresses on many other occasions, among which may be mentioned a dinner to Hógg the Ettrick Shepherd at Peebles
(where he captivated Professor Wilson); various celebrations
in connection with the Buccleuch family (with whom he had
long been held in high regard); the inauguration of the
Ettrick Shepherd's monument at St Mary's Loch; and the
celebration of Sir Walter Scott's centenary at Branxholm,
which was his last appearance before a numerous promiscuous
audience.

His repute as a breeder of sheep, his intellectual endowments,
his sagacity and independence of judgment, his frank and
obliging nature, and his gentlemanly bearing all combined to
concentrate in him the admiration and confidence of the com-
munity, which grew and deepened with his years.   These
qualities made him a favourite arbiter, in which capacity he
often acted; and they also brought upon him frequent requests
to officiate as judge at the great national Agricultural as well
as local societies' shows.   It need not excite surprise that such
a man should be chosen president of the Teviotdale Agricultural Society at its formation in 1859, and that, with the exception of one year, when he retired in favour of a friend, he

should have been annually re-elected to that office till his removal by death. So highly did the members esteem the man and value his services, that they had his portrait painted and hung up in their hall in the Tower Hotel at Hawick, the occasion of its being unveiled being taken to honour Mr. Aitchison by entertaining him to a banquet. As a man of business he was prompt and diligent, careful and yet generous in all his dealings, a kind and considerate master, and everything he did was done most pleasantly. Ever ready to lend a helping hand by counsel or assistance to those who required and desired it, many a young farmer can date the era of his prosperity from the time Mr. Aitchison kindly interested himself in his welfare. He had great discernment of character, and holding in utter scorn all that was mean and dishonest he could rarely be deceived by the unworthy, so his favours were dispensed most prudently. Highly endowed with that tact which enables a man to secure the confidence of his fellows, he had the faculty of suiting his bearing and conversation to all orders of men with whom he came in contact, for he uniformly mingled dignity with courtesy. His excellent social qualities made his society courted, and there was a great charm in his conversation, for he had read much, studied men keenly, and had an immense fund of anecdote. He had a fine humorous vein, and could alternately call down thunders of applause, and set the table in a roar with bright flashes of wit and raillery. No man on the Border could be pointed to as more truly a representative man of his class than Mr. Aitchison, and the intimation of his death was received with no common grief.

His remains were deposited in the burying-ground at Unthank, Ewes, where his ancestors have been interred since the beginning of the eighteenth century, and the funeral was attended by such a train of mourners as is only to be seen in the funeral cortege of men of unusual mark and merit.

Improvement of the Blackfaced Breed.

The following remarks are extracted from a paper, published in 1792, from the pen of Sir John Sinclair, Bart., who was then Chairman of the Society for the Improvement of British Wool, and who took great interest in the improvement of the different breeds of Sheep in this country :—

" Perhaps there is no part of the whole island where, at first sight, a fine-woolled breed of sheep is less to be expected than among the Cheviot hills. Many parts of the sheep walks there consist of nothing but peat bogs and deep morasses. During the winter the hills are covered with snow for two, three, and sometimes even four months, and they have an ample proportion of bad weather during the other seasons of the year; yet there a species of sheep are to be found, taking all their properties together, equal, if not superior, to any other in Great Britain *for a mountainous district*, and which will thrive even in the wildest parts of it.

" These sheep are long bodied; they have, in general, thirteen, but sometimes fourteen, ribs on each side; their shape is excellent, and their fore quarters, in particular, are distinguished by such justness of proportion as to be equal in weight to the hind. Their limbs are of a length to fit them for travelling, and to enable them to pass over bogs and snows, through which a shorter legged animal could not well penetrate; they are polled, white faced, and have rarely any black spots on any part of their body; they have a closer fleece than the Tweeddale, or Linton breed, which keeps them warmer in cold weather, and prevents either rain or snow from incommoding them; their fleece is shorter, and, of consequence, it is evidently more portable over mountainous pastures; they are excellent snow breakers, and are accustomed to procure their food by scraping the snow off the ground with their feet, even when the top is hardened by frost; they have never any other food (unless when it is proposed to fatten them) besides the grass and natural hay produced on their own hills. They are, it is said, less subject to diseases than the common blackfaced kind, particularly to what is called the *braxy* or the *sickness*. They sell at a good price to the grazier, and their value for feeding is rising every day; the draft or cast ewes, when lean, now fetching from 12s to 16s a-piece, and three-year-old wedders from 18s to

F

22s. Their weight when fat, at four years old, is from 17 to 20 lbs. per quarter; and the mutton, when fed upon heath, and kept to a proper age, is fully equal, in taste or flavour, to any that the Highlands produce. Lambs fed for the butcher, on the milk of the ewe, now fetch from 8s to 10s a-piece. From eight to nine fleeces of white wool make a stone of 24 lbs. weight, and from six to eight fleeces when the wool is laid or smeared. The laid or smeared wool sold, in 1792, at from 18s to 20s, and the white from 20s to 22s. Some went as high as 23s, and from the improvements now going forward, it will soon fetch 30s, if not 40s per stone. Their superiority over the Tweeddale or common blackfaced breed is incontrovertibly proved by a variety of experiments. Mr Thomas Scott at Lethem, on Carter Fell, a mountain about 1600 feet above the level of the sea, exchanged, in 1773, with Walter Hog in Ettrick forest, five whitefaced for as many blackfaced tups, but had every reason to regret the experiment, which was far from being the case with Mr Hog. Mr Roger Marshall at Blindburn, in Northumberland, came to that farm in 1769, and purchased the stock upon the ground, among which there were many blackfaced sheep. These he completely extirpated, and found it greatly to his advantage. So much convinced, indeed, are the farmers in the neighbourhood—particularly those of Ettrick forest, of Ewesdale, and Liddesdale—of their superior excellence, that they are now converting their flocks, as quickly as possible, into the Cheviot breed.

" The progress that has been made in improving this breed, particularly in regard to meliorating its wool, is in the highest degree satisfactory. About twenty years ago the average of the white and laid wool was about ten fleeces to the stone, which sold for about 8s ; whereas now the value is not only more than doubled, but the weight is so much increased that, at an average, eight fleeces will make a stone. Even this excellent breed, however, is still capable of some improvement, and the experiments which are now going forward will soon ascertain by what means that improvement can best be procured. The shape of the animal, for a hilly district, is almost brought to perfection ; but the wool requires—first, to be still finer in the pile ; secondly, shorter in the staple, so as to make it not only more portable for the animal itself, but fitter for being manufactured into cloth ; thirdly, thicker in the coat, so as to keep the animal warmer ; and lastly, more equal in point of quality *(a circumstance of very great importance)*, so that the whole fleece may be as nearly as possible the same. These are qualities which the Spanish, the Hereford, and the Southdown breeds possess in very great perfection ; and if once the hardiness, the excellent carcase, and the other advantages of the Cheviot breed were united to these properties, *hill sheep would be brought to their greatest height of perfection.* All these different crosses are now under trial ; and, as yet, every one of them seems to have succeeded, so that any of them may be followed with success. In regard to the original Cheviot breed, they have been tried, on a greater or lesser scale, in every part of Scotland; and on every occasion they have answered in the wildest parts of the country, and even in places where no sheep were ever kept before, at least in any quantity."

www.ingramcontent.com/pod-product-compliance
Lightning Source LLC
Chambersburg PA
CBHW081742220526
45468CB00008B/2203